自贡恐龙博物馆

馆藏精品

ZIGONG DINOSAUR MUSEUM
TREASURES

自贡恐龙博物馆 编

科学出版社

北京

内 容 简 介

自贡恐龙博物馆是中国第一座专业性恐龙博物馆和世界上著名的恐龙博物馆之一。博物馆拥有丰富的恐龙化石及其他古生物化石，其中有众多的世界级珍品。本书从丰富的馆藏标本中精选了49件代表性恐龙化石和37件恐龙时代的其他脊椎动物化石编辑成册，不仅展现了博物馆丰富多样的古生物化石收藏，也能让广大读者领略亿万年前古生物化石的神奇魅力。

本书可供广大地质古生物工作者、博物馆从业人员和古生物爱好者参考，同时也是开展古生物化石科普教育和宣传的极好素材。

图书在版编目(CIP)数据

自贡恐龙博物馆馆藏精品 / 自贡恐龙博物馆编. —北京：科学出版社，2019.8
ISBN 978-7-03-062032-3

Ⅰ. ①自⋯　Ⅱ. ①自⋯　Ⅲ. ①恐龙化石-图集　Ⅳ. ①Q915.2-64

中国版本图书馆CIP数据核字（2019）第169028号

责任编辑：罗　莉 / 责任校对：彭　映
责任印制：罗　科 / 封面设计：凌　曼

科学出版社 出版
北京东黄城根北街16号
邮政编码：100717
http://www.sciencep.com

四川煤田地质制图印刷厂 印刷
科学出版社发行　各地新华书店经销

*

2019年8月第　一　版　　开本：889×1194　1/16
2019年8月第一次印刷　　印张：8
字数：200 000

定价：120.00元
（如有印装质量问题，我社负责调换）

编委会

主　　编：李　健
副 主 编：彭光照　李　飚
编　　委：（按姓氏笔画排序）
　　　　　王　波　叶　勇　江　山　余　刚　余　勇　罗　舒　胡晓冬　郝宝鞘　凌　曼
编　　辑：叶　勇
摄　　影：余　刚　王　波　陈　立　余　勇
设　　计：余　勇　凌　曼　罗　舒
翻　　译：李　毅

序

记得四十年前，为完成中英联合恐龙考察选点任务，我带考察小组来到自贡大山铺。一个震惊世界的恐龙化石宝库的大门，随着四川石油管理局川西南矿区的工地施工而被叩开。伴随着大山铺恐龙化石的大规模清理和发掘，数以万计的恐龙骨骼化石被发现，这个化石宝库的神秘面纱也逐渐被揭开。三十多年前，中国第一座专业性的恐龙遗址博物馆——自贡恐龙博物馆的建立，把我国恐龙的发掘、研究工作推向了一个新的阶段，成为我国恐龙研究史上的一个新的里程碑。自贡这个有着千余年井盐生产历史的城市，也成为了中国恐龙学者云集的地方和国际恐龙学界人士必须朝拜的"圣地"。作为自贡恐龙的发现和研究者，我感到无比骄傲和自豪。

藏品是博物馆的基础，自贡恐龙博物馆自建立以来就非常重视藏品的收藏与研究。作为目前世界上拥有最大规模恐龙化石埋藏现场、展示和收藏中侏罗世恐龙化石最丰富的博物馆，它拥有许多世界级和国家级的馆藏精品，如世界上所知最原始、保存最完整的剑龙化石——太白华阳龙化石；世界上首例剑龙类皮肤（印痕）化石——四川巨棘龙皮肤（印痕）化石；我国首例蜥脚类恐龙皮肤（印痕）化石——杨氏马门溪龙皮肤（印痕）化石；世界上首次发现与肩带骨骼呈关节状态保存的肩棘化石——四川巨棘龙肩棘化石；世界上首次发现的蜥脚类恐龙骨质尾锤化石——李氏蜀龙尾锤化石；等等。这些精品化石不仅具有重要的科学研究价值，而且具有很高的科普观赏价值。

在中华人民共和国成立70周年和自贡市设市80周年之际，本书的编者们精心甄选了部分馆藏精品图片辑成这本图录。这是自贡恐龙博物馆馆藏精品的第一次集中展示，同时为我们提供了一本非常有价值的参考资料，在此特表示祝贺。我也衷心希望：自贡恐龙博物馆在繁荣中国恐龙事业的道路上展翅飞翔、再铸辉煌。

2019年7月2日于云南禄丰

FORWORD

ZIGONG
DINOSAUR MUSEUM
TREASURES

Forty years ago, in order to find potential sites for the Sino-UK Dinosaur Expedition, my research team and I came to Dashanpu, Zigong, where a world-shaking treasure house of dinosaur fossils was being opened during the construction of the Southwest Petroleum Mining. With extensive excavations in the following years, tens of thousands of dinosaur bones were uncovered and the mystery of this dinosaur quarry was gradually unveiled. More than thirty years ago, the foundation of the Zigong Dinosaur Museum, being a new milestone in the history of dinosaur research, promoted the dinosaur research in China to a new stage. And, city of Zigong, with a history of well-salt production industry over thousand years, attracted a lot of Chinese paleontologists, and has become one of the places the dinosaur experts all over the world desire to visit. I am very proud that I discovered and have worked on the dinosaurs of Zigong.

Collections are the basis of a museum. Since its establishment, the Zigong Dinosaur Museum has put lots of effort on both collections and researches. With the biggest burial site and the most abundant collection and exhibition of dinosaurs from the Middle Jurassic in the world, the museum holds a great number of world- and national-class specimens, such as *Huayangosaurus taibaii*, the most primitive and best-preserved stegosaur so far known in the world; the skin (impression) fossil of *Gigantspinosaurus sichuanensis*, the first skin (impression) fossil of stegosaurs in the world; the skin (impression) fossil of *Mamenchisaurus youngi*, the first sauropod skin (impression) fossil in China; the parascapular spines of *Gigantspinosaurus sichuanensis*, the first parascapular spine articulated with the shoulder bones in the world; the tail club of *Shunosaurus lii*, the first bony club of sauropods in the world. These delicate fossils are of great scientific values.

On the occasions of the 70th anniversary of the foundation of the People's Republic of China and the 80th anniversary of the establishment of Zigong City, the editors elaborately prepared this selected catalogue based on the rare and delicate specimens in the museum. It's the first time to collectively display these high-quality collections of the Zigong Dinosaur Museum, and will be a valuable reference. I would like to extend my congratulations to the editors, and hope the museum will make more breakthroughs in the future.

Dong Zhiming
July 2, 2019
Lufeng, Yunnan

自贡恐龙博物馆馆藏精品

前 言

自贡恐龙博物馆建立于举世闻名的大山铺恐龙化石群遗址之上，是中国第一座专业性恐龙博物馆、首批国家一级博物馆和自贡世界地质公园的核心园区，被美国《国家地理》评价为"世界上最好的恐龙博物馆"。历经三十多年的发展和几代博物馆人的努力，博物馆的化石收藏越来越丰富，其中不乏众多的世界级珍品。值此建国70周年和自贡设市80周年之际，我们编辑出版了博物馆的第一本馆藏精品图录，谨此献给我们伟大的祖国和美丽的家乡，献给为自贡恐龙化石的发现和保护、为自贡恐龙博物馆的建设和发展作出过贡献的人们。

本书从博物馆众多的馆藏标本中精选了49件恐龙化石和37件其他脊椎动物化石，种类丰富多样：不仅有太白华阳龙、李氏蜀龙等众多保存完整的恐龙骨架和李氏蜀龙尾锤、太白华阳龙肩棘等独具自贡特色的典型恐龙骨骼化石，而且有保存精美的恐龙皮肤（印痕）、恐龙足迹和窝状恐龙蛋等珍稀恐龙遗迹化石，同时还有鱼类、两栖类、龟类、鳄类、蛇颈龙类、翼龙类和似哺乳爬行类等恐龙时代不同门类的伴生脊椎动物化石，从而让广大读者领略亿万年前古生物化石的神奇魅力。

本书是集体智慧的结晶，是自贡恐龙博物馆众多同仁团结协作的成果。我们不会忘记：为了保证恐龙骨架照片的质量，我们的科研人员、摄影师和技术工人们在户外顶着烈日对多具大型恐龙骨架进行了重新组装和拍摄；为了获得更好的创意和设计效果，我们的美术设计师对图录封面和内页设计进行了多次修改完善；为了保证内容的科学性，我们的恐龙专家对文字部分，特别是英文和拉丁文进行了一遍又一遍的仔细检查校对。在此，对他们的辛勤付出表示衷心的感谢！同时也对关心、支持和协助本书编纂和出版工作的领导、同行及科学出版社的同仁们表示衷心的感谢！

PREFACE

Zigong Dinosur Museaum is built in situ right at the world-famous Dashanpu fossil site. It is the first professional dinosaur museum in China, and one of the first National First-class Museums. Being the core area of Zigong Global Geopark, the museum was evaluated as "the best dinosaur museum in the world" by *National Geographic* of the United States. With the development of more than 30 years and efforts of several generations of the museum staff, the collections have been increasingly accumulated, among which are many world-class treasures. On the occasions of the 70th anniversary of the foundation of the People's Republic of China and the 80th anniversary of the establishment of Zigong City, we made the first selected catalogue of Zigong Dinosaur Museum to dedicate to our great motherland and beautiful hometown, and to the people who have contributed to the discovery and protection of Zigong dinosaur fossils and the establishment and development of the museum.

49 dinosaurs and 37 other vertebrates of various species have been selected from numerous collections of the museum. There are many well-preserved dinosaur skeletons, including *Huayangosaurus taibaii*, and *Shuonosaurus lii*, as well as the unique specimens from Zigong, such as the tail club of *Shuonosaurus lii*, spine of *Huayangosaurus taibaii*. There are also rare but exquisitely preserved impression of dinosaur skin, dinosaur footprints and eggs. All these dinosaur fossils, along with various kinds of associated vertebrates such as fishes, amphibians, turtles, crocodiles, plesiosaurs, pterosaurs and mammal-like reptiles, present the miraculous beauty and glamour of the ancient animals from hundreds of millions years ago.

This book is a collective piece of work made by a extensive collaboration of all the colleagues in the museum. We will never forget, in order to get the high-quality pictures, our researchers and technicians reassembled many large skeletons and our photographers took photos under the scorching sun-light; to obtain a better creative idea and the best design, our art designers revised and polishes the cover and the layout of each page again and again; our dinosaur experts read the text throughly and carefully over and over, especially the English and Latin parts, to assure scientific accuracy of the book. Herein, we would like to wholeheartedly express our gratitude for the hard and great work they have done. At the meantime, many thanks go to the leaders, and colleagues from the Science Press and other presses who have been supportive and helpful to get this book published!

目 录 CONTENTS

恐龙化石精品 DINOSAUR TREASURES ... 1

太白华阳龙 *Huayangosaurus taibaii* ... 2
四川巨棘龙 *Gigantspinosaurus sichuanensis* ... 4
李氏蜀龙 *Shunosaurus lii* ... 6
天府峨眉龙 *Omeisaurus tianfuensis* ... 10
杨氏马门溪龙 *Mamenchisaurus youngi* ... 14
合川马门溪龙 *Mamenchisaurus hochuanensis* ... 16
巴山酋龙 *Datousaurus bashanensis* ... 18
董氏大山铺龙 *Dashanpusaurus dongi* ... 20
和平永川龙 *Yangchuanosaurus hepingensis* ... 22
自贡四川龙 *Szechuanosaurus zigongensis* ... 24
多齿何信禄龙 *Hexinlusaurus multidens* ... 26
劳氏灵龙 *Agilisaurus louderbacki* ... 28
中国猎龙（未定种）*Sinovenator* sp. ... 30
近鸟龙（未定种）*Anciornis* sp. ... 31
和平永川龙头骨 skull of *Yangchuanosaurus hepingensis* ... 32
东坡秀龙头骨 skull of *Abrosaurus dongpoi* ... 34
李氏蜀龙头骨 skull of *Shunosaurus lii* ... 35
杨氏马门溪龙头骨 skull of *Mamenchisaurus youngi* ... 38
巴山酋龙头骨 skull of *Datousaurus bashanensis* ... 40
四川巨棘龙下颌 mandible of *Gigantspinosaurus sichuanensis* ... 41
太白华阳龙头骨 skull of *Huayangosaurus taibaii* ... 42
劳氏灵龙头骨 skull of *Agilisaurus louderbacki* ... 44
多齿何信禄龙头骨 skull of *Hexinlusaurus multidens* ... 45
合川马门溪龙头骨 skull of *Mamenchisaurus hochuanensis* ... 46
自贡四川龙下颌 mandible of *Szechuanosaurus zigongensis* ... 47
杨氏马门溪龙左前脚 left manus of *Mamenchisaurus youngi* ... 48
杨氏马门溪龙皮肤（印痕）skin impression of *Mamenchisaurus youngi* ... 49
四川巨棘龙肩棘和皮肤（印痕）parascapular spine and skin impression of *Gigantspinosaurus sichuanensis* ... 50
太白华阳龙肩棘 parascapular spine of *Huayangosaurus taibaii* ... 52
李氏蜀龙尾锤 tail club of *Shunosaurus lii* ... 54
天府峨眉龙尾锤 tail club of *Omeisaurus tianfuensis* ... 56
扁圆蛋（未定种）*Placoolithus* sp. ... 60
长形蛋（未定种）*Elongatoolithus* sp. ... 62
巨型长形蛋（未定种）*Macroelongatoolithus* sp. ... 64

自贡实雷龙足迹 *Eubrontes zigongensis* 66
何氏极大龙足迹 *Gigandipus hei* 67
兽脚类足迹和蜥脚类幻迹 footprints of theropods and transmitted prints of sauropods 68
跷脚龙类足迹 footprints of grallatorids 72

伴生动物化石精品 COEXISTENT VERTEBRATE TREASURES 74

大山铺鳞齿鱼 *Lepidotes dashanpuensis* 76
刘氏原白鲟 *Protopsephurus liui* 78
自贡角齿鱼齿板 dental plate of *Ceratodus zigongensis* 80
扁头中国短头鲵头骨 skull of *Sinobrachyops placenticephalus* 81
自贡巴蜀龟 *Bashuchelys zigongensis* 82
大山铺川南龟 *Chuannanchelys dashanpuensis* 82
自贡天府龟 *Tienfuchelys zigongensis* 83
杨氏巴蜀龟 *Bashuchelys youngi* 84
周氏四川龟 *Sichuanchelys chowi* 86
盐原新疆龟 *Protoxinjiangchelys salis* 88
叶氏香港龟 *Hongkongochelys yehi* 89
放射纹成渝龟 *Chengyuchelys radiplicatus* 90
宽缘板成渝龟（相似种）*Chengyuchelys* cf. *latimarginalis* 91
娇小盐都龟 *Yanduchelys delicatus* 92
重庆天府龟（相似种）*Tienfuchelys* cf. *chungkingensis* 93
原新疆龟（相似属）cf. *Protoxinjiangchelys* 94
南雄龟属（未定种）*Nanhsiungchelys* sp. 95
汇东四川鳄 *Sichuanosuchus huidongensis* 96
蜀南孙氏鳄头骨 skull of *Sunosuchus shunanensis* 98
周氏西蜀鳄头骨 skull of *Hsisosuchus chowi* 99
大山铺西蜀鳄头骨 skull of *Hsisosuchus dashanpuensis* 100
沱江西蜀鳄头骨 skull of *Hsisosuchus tuojiangensis* 101
沱江西蜀鳄 *Hsisosuchus tuojiangensis* 102
自贡璧山上龙 *Bishanopliosaurus zigongensis* 104
贵州鱼龙（未定种）*Guizhouichthyosaurus* sp. 106
凌源潜龙 *Hyphalosaurus lingyuanensis* 108
长头狭鼻翼龙头骨 skull of *Angustinaripterus longicephalus* 110
董氏中国翼龙 *Sinopterus dongi* 111
川南多齿兽头骨 skull of *Polistodon chuannanensis* 112
爬兽（未定种）头骨 skull of *Repenomamus* sp. 113
孤反鸟（未定种）*Monoenantiornis* sp. 114
朝阳鸟（未定种）*Chaoyangia* sp. 115
手兽足迹（未定种）*Chirotherium* sp. 116

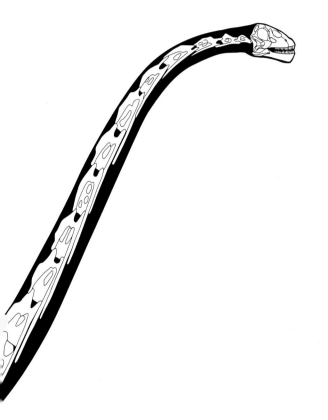

恐龙 化石精品

DINOSAUR TREASURES

1 ▶

恐龙化石精品

	2.01亿年	1.45亿年		0.66亿年
三 叠 纪	侏 罗 纪	白 垩 纪		化石的地质年代
T₁　T₂　T₃	J₁　J₂　J₃	K₁	K₂	（全书同）

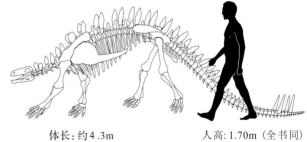

体长：约4.3m　　　　　人高：1.70m（全书同）

太白华阳龙
Huayangosaurus taibaii

类别：剑龙亚目剑龙科
产地：自贡大山铺
Classification: Stegosauridae,
　　　　　　　Stegoasuria
Locality: Dashanpu, Zigong

恐龙化石精品

2.52亿年	2.01亿年	1.45亿年	0.66亿年
三叠纪	侏罗纪	白垩纪	

自贡恐龙博物馆 馆藏精品

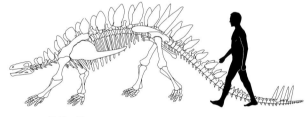

体长：约5.4 m

四川巨棘龙
Gigantspinosaurus sichuanensis

类别：剑龙亚目剑龙科
产地：自贡仲权

Classification: Stegosauridae,
　　　　　　　Stegoasuria
Locality: Zhongquan, Zigong

5

恐龙化石精品

2亿年	2.01亿年	1.45亿年	0.66亿年
三叠纪	侏罗纪	白垩纪	

自贡恐龙博物馆 馆藏精品

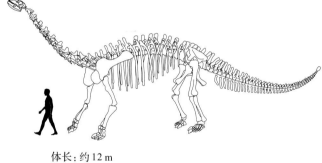

体长：约12 m

李氏蜀龙
Shunosaurus lii

类别：蜥脚形亚目妖龙科
产地：自贡大山铺

Classification: Cetiosauridae, Sauropodomorpha
Locality: Dashanpu, Zigong

恐龙化石精品

| 52亿年 | 2.01亿年 | 1.45亿年 | 0.66亿年 |

| 三叠纪 | 侏罗纪 | 白垩纪 |

馆藏精品

体长：约6 m

李氏蜀龙
Shunosaurus lii

类别：蜥脚形亚目妖龙科
产地：自贡大山铺
Classification: Cetiosauridae, Sauropodomorpha
Locality: Dashanpu, Zigong

52亿年	2.01亿年	1.45亿年	0.66亿年
三叠纪	侏罗纪	白垩纪	

体长：约20 m

天府峨眉龙
Omeisaurus tianfuensis

类别：蜥脚形亚目马门溪龙科
产地：自贡大山铺

Classification: Mamenchisauridae, Sauropodomorpha
Locality: Dashanpu, Zigong

恐龙化石精品

2.52亿年	2.01亿年	1.45亿年	0.66亿年
三叠纪	侏罗纪	白垩纪	

体长：约6 m

天府峨眉龙
Omeisaurus tianfuensis

类别：蜥脚形亚目马门溪龙科
产地：自贡大山铺
Classification: Mamenchisauridae, Sauropodomorpha
Locality: Dashanpu, Zigong

恐龙化石精品

杨氏马门溪龙
Mamenchisaurus youngi

类别：蜥脚形亚目马门溪龙科
产地：自贡新民

Classification: Mamenchisauridae, Sauropodomorpha
Locality: Xinmin, Zigong

2.52亿年	2.01亿年	1.45亿年	0.66亿年
三叠纪	侏罗纪	白垩纪	

自贡恐龙博物馆

馆藏精品

体长：约17 m

15

恐龙化石精品

| 2亿年 | 2.01亿年 | 1.45亿年 | 0.66亿年 |

| 三叠纪 | 侏罗纪 | 白垩纪 |

体长：约21 m

合川马门溪龙
Mamenchisaurus hochuanensis

类别：蜥脚形亚目马门溪龙科
产地：自贡汇东

Classification: Mamenchisauridae, Sauropodomorpha
Locality: Huidong, Zigong

恐龙化石精品

巴山酋龙
Datousaurus bashanensis

类别：蜥脚形亚目圆顶龙科
产地：自贡大山铺

Classification: Camarasauridae, Sauropodomorpha
Locality: Dashanpu, Zigong

| 2.52亿年 | 2.01亿年 | 1.45亿年 | 0.66亿年 |
| 三叠纪 | 侏罗纪 | 白垩纪 |

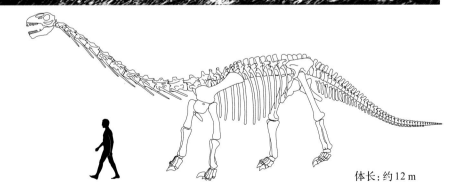

体长：约12 m

董氏大山铺龙
Dashanpusaurus dongi

类别：蜥脚形亚目圆顶龙科
产地：自贡大山铺

Classification: Camarasauridae, Sauropodomorpha
Locality: Dashanpu, Zigong

2.52亿年	2.01亿年	1.45亿年	0.66亿年
三叠纪	侏罗纪	白垩纪	

馆藏精品

自贡恐龙博物馆

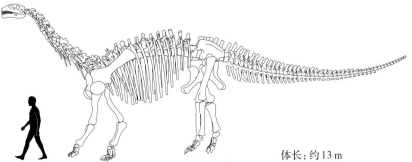

体长：约13 m

■ 恐龙化石精品

2.52亿年	2.01亿年	1.45亿年	0.66亿年
三叠纪	侏罗纪	白垩纪	

体长：约9.0 m

和平永川龙
Yangchuanosaurus hepingensis

类别：兽脚亚目巨齿龙科
产地：自贡和平

Classification: Megalosauridae, Theropoda
Locality: Heping, Zigong

■ 恐龙化石精品

自贡四川龙
Szechuanosaurus zigongensis

类别：兽脚亚目巨齿龙科
产地：自贡大山铺

Classification: Megalosauridae, Theropoda
Locality: Dashanpu, Zigong

2.52亿年	2.01亿年	1.45亿年	0.66亿年
三叠纪	侏罗纪	白垩纪	

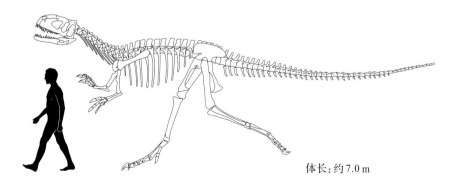

体长：约7.0 m

馆藏精品

恐龙化石精品

多齿何信禄龙
Hexinlusaurus multidens

类别：鸟脚亚目法布劳龙科
产地：自贡大山铺

Classification: Fabrosauridae, Ornithopoda
Locality: Dashanpu, Zigong

2.52亿年	2.01亿年	1.45亿年	0.66亿年
三叠纪	侏罗纪	白垩纪	

自贡恐龙博物馆 馆藏精品

体长：约1.4 m

27

恐龙化石精品

劳氏灵龙
Agilisaurus louderbacki

类别：鸟脚亚目法布劳龙科
产地：自贡大山铺

Classification: Fabrosauridae, Ornithopoda
Locality: Dashanpu, Zigong

2.52亿年	2.01亿年	1.45亿年	0.66亿年
三叠纪	侏罗纪	白垩纪	

馆藏精品

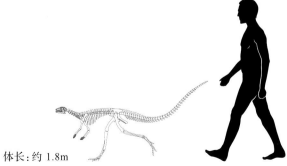

体长：约1.8m

29

恐龙化石精品

2.52亿年	2.01亿年	1.45亿年	0.66亿年
三叠纪	侏罗纪	白垩纪	

中国猎龙（未定种）
Sinovenator sp.

类别：兽脚亚目伤齿龙科
产地：辽宁北票

Classification: Troodontidae, Theropoda
Locality: Beipiao, Liaoning

2.52亿年	2.01亿年	1.45亿年	0.66亿年
三叠纪	侏罗纪	白垩纪	

近鸟龙（未定种）
Anciornis sp.

类别：兽脚亚目伤齿龙科
产地：辽宁
Classification: Troodontidae, Theropoda
Locality: Liaoning

■ 恐龙化石精品

| 2.52亿年 | 2.01亿年 | 1.45亿年 | 0.66亿年 |

| 三叠纪 | 侏罗纪 | 白垩纪 |

自贡恐龙博物馆 馆藏精品

和平永川龙头骨
skull of
Yangchuanosaurus hepingensis

类别：兽脚亚目巨齿龙科
产地：自贡和平
Classification: Megalosauridae, Theropoda
Locality: Heping, Zigong

恐龙化石精品

	2.52亿年	2.01亿年	1.45亿年	0.66亿年
	三叠纪	侏罗纪	白垩纪	

东坡秀龙头骨
skull of *Abrosaurus dongpoi*

类别：蜥脚形亚目马门溪龙科
产地：自贡大山铺

Classification: Mamenchisauridae, Sauropodomorpha
Locality: Dashanpu, Zigong

2.52亿年　　　2.01亿年　　　1.45亿年　　　　　　　　0.66亿年

三叠纪　　侏罗纪　　　白垩纪

李氏蜀龙头骨
skull of *Shunosaurus lii*

类别：蜥脚形亚目妖龙科
产地：自贡大山铺

Classification: Cetiosauridae, Sauropodomorpha
Locality: Dashanpu, Zigong

恐龙化石精品

| 2.52亿年 | 2.01亿年 | 1.45亿年 | 0.66亿年 |

| 三叠纪 | 侏罗纪 | 白垩纪 |

李氏蜀龙头骨
skull of *Shunosaurus lii*

类别：蜥脚形亚目妖龙科
产地：自贡大山铺

Classification: Cetiosauridae, Sauropodomorpha
Locality: Dashanpu, Zigong

自贡恐龙博物馆 馆藏精品

恐龙化石精品

2.52亿年	2.01亿年	1.45亿年	0.66亿年
三叠纪	侏罗纪	白垩纪	

杨氏马门溪龙头骨
skull of *Mamenchisaurus youngi*

类别：蜥脚形亚目马门溪龙科
产地：自贡新民

Classification: Mamenchisauridae, Sauropodomorpha
Locality: Xinmin, Zigong

巴山酋龙头骨
skull of *Datousaurus bashanensis*

类别：蜥脚形亚目圆顶龙科
产地：自贡大山铺

Classification: Camarasauridae, Sauropodomorpha
Locality: Dashanpu, Zigong

四川巨棘龙下颌
mandible of *Gigantspinosaurus sichuanensis*

类别：剑龙亚目剑龙科
产地：自贡仲权

Classification: Stegosauridae, Stegoasuria
Locality: Zhongquan, Zigong

恐龙化石精品

.52亿年	2.01亿年	1.45亿年	0.66亿年
三叠纪	侏罗纪	白垩纪	

自贡恐龙博物馆 馆藏精品

太白华阳龙头骨
skull of *Huayangosaurus taibaii*

类别：剑龙亚目剑龙科
产地：自贡大山铺

Classification: Stegosauridae,
　　　　　　　Stegoasuria
Locality: Dashanpu, Zigong

43

恐龙化石精品

劳氏灵龙头骨
skull of *Agilisaurus louderbacki*

类别：鸟脚亚目法布劳龙科
产地：自贡大山铺

Classification: Fabrosauridae, Ornithopoda
Locality: Dashanpu, Zigong

2.52亿年	2.01亿年	1.45亿年	0.66亿年
三叠纪	侏罗纪	白垩纪	

多齿何信禄龙头骨
skull of *Hexinlusaurus multidens*

类别：鸟脚亚目法布劳龙科
产地：自贡大山铺

Classification: Fabrosauridae, Ornithopoda
Locality: Dashanpu, Zigong

恐龙化石精品

合川马门溪龙头骨
skull of *Mamenchisaurus hochuanensis*

类别：蜥脚形亚目马门溪龙科
产地：自贡汇东

Classification: Mamenchisauridae, Sauropodomorpha
Locality: Huidong, Zigong

自贡四川龙下颌
mandible of *Szechuanosaurus zigongensis*

类别：兽脚亚目巨齿龙科
产地：自贡大山铺

Classification: Megalosauridae, Theropoda
Locality: Dashanpu, Zigong

恐龙化石精品

| 2.52亿年 | 2.01亿年 | 1.45亿年 | 0.66亿年 |
| 三叠纪 | 侏罗纪 | 白垩纪 |

杨氏马门溪龙左前脚
left manus of *Mamenchisaurus youngi*

类别：蜥脚形亚目马门溪龙科
产地：自贡新民

Classification: Mamenchisauridae, Sauropodomorpha
Locality: Xinmin, Zigong

	2.52亿年	2.01亿年	1.45亿年	0.66亿年
	三叠纪	侏罗纪	白垩纪	

）痕）
enchisaurus youngi

奚龙科

dae, Sauropodomorpha

恐龙化石精品

50

四川巨棘龙肩棘和皮肤（印痕）
parascapular spine and skin impression of *Gigantspinosaurus sichuanensis*

类别：剑龙亚目剑龙科
产地：自贡仲权

Classification: Stegosauridae, Stegoasuria
Locality: Zhongquan, Zigong

■ 恐龙化石精品

| 2.52亿年 | 2.01亿年 | 1.45亿年 | 0.66亿年 |

| 三叠纪 | 侏罗纪 | 白垩纪 |

太白华阳龙肩棘
parascapular spine of
Huayangosaurus taibaii

类别：剑龙亚目剑龙科
产地：自贡大山铺
Classification: Stegosauridae,
　　　　　　　Stegoasuria
Locality: Dashanpu, Zigong

恐龙化石精品

	2.01亿年	1.45亿年	0.66亿年
三叠纪	侏罗纪	白垩纪	

李氏蜀龙尾锤
tail club of *Shunosaurus lii*

类别：蜥脚形亚目妖龙科
产地：自贡大山铺

Classification: Cetiosauridae, Sauropodomorpha
Locality: Dashanpu, Zigong

■ 恐龙化石精品

| 2.52亿年 | 2.01亿年 | 1.45亿年 | 0.66亿年 |

| 三叠纪 | 侏罗纪 | 白垩纪 |

天府峨眉龙尾锤
tail club of *Omeisaurus tianfuensis*

类别：蜥脚形亚目马门溪龙科
产地：自贡大山铺

Classification: Mamenchisauridae, Sauropodomorpha
Locality: Dashanpu, Zigong

2.52亿年　　　2.01亿年　　　1.45亿年　　　　　　0.66亿年

| 三叠纪 | 侏罗纪 | 白垩纪 |

天府峨眉龙尾锤
tail club of *Omeisaurus tianfuensis*

类别：蜥脚形亚目马门溪龙科
产地：自贡大山铺

Classification: Mamenchisauridae, Sauropodomorpha
Locality: Dashanpu, Zigong

恐龙化石精品

天府峨眉龙尾锤
tail club of *Omeisaurus tianfuensis*

类别：蜥脚形亚目马门溪龙科
产地：自贡大山铺
Classification: Mamenchisauridae, Sauropodomorpha
Locality: Dashanpu, Zigong

天府峨眉龙尾锤
tail club of *Omeisaurus tianfuensis*

类别：蜥脚形亚目马门溪龙科
产地：自贡大山铺

Classification: Mamenchisauridae, Sauropodomorpha
Locality: Dashanpu, Zigong

■ 恐龙化石精品

扁圆蛋（未定种）
Placoolithus sp.

类别：树枝蛋科
产地：河南西峡

Classification: Dendroolithidae
Locality: Xixia, Henan

2.52亿年	2.01亿年	1.45亿年	0.66亿年
三叠纪	侏罗纪	白垩纪	

馆藏精品

恐龙化石精品

长形蛋（未定种）
Elongatoolithus sp.

类别：长形蛋科
产地：广东南雄
Classification: Elongatoolithidae
Locality: Nanxiong, Guangdong

2.52亿年	2.01亿年	1.45亿年	0.66亿年
三叠纪	侏罗纪	白垩纪	

长形蛋（未定种）
Elongatoolithus sp.

类别：长形蛋科
产地：广东南雄
Classification: Elongatoolithidae
Locality: Nanxiong, Guangdong

恐龙化石精品

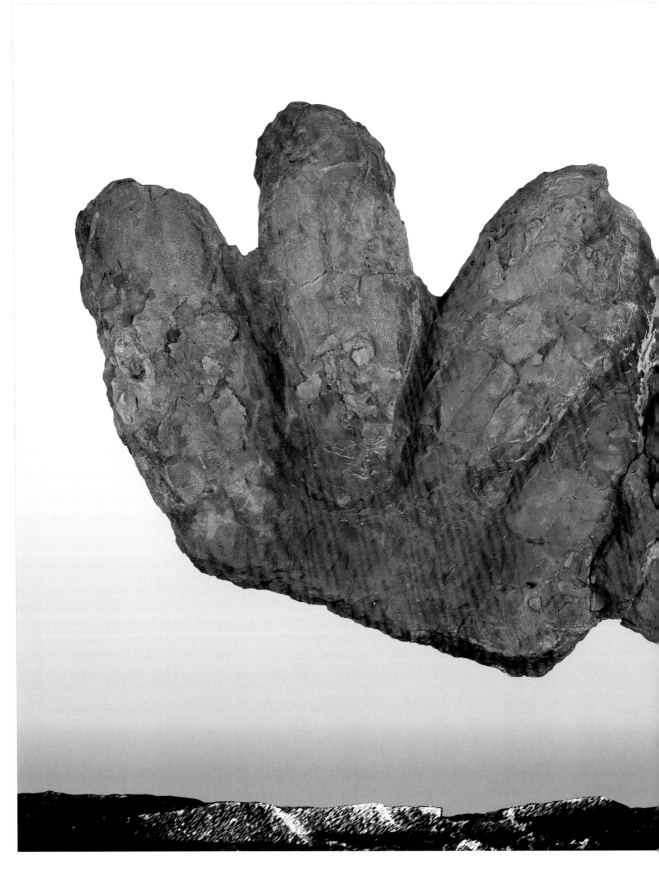

2.52亿年	2.01亿年	1.45亿年	0.66亿年
三叠纪	侏罗纪	白垩纪	

巨型长形蛋（未定种）
Macroelongatoolithus sp.

类别：巨型长形蛋科
产地：河南西峡

Classification: Macroelongatoolithidae
Locality: Xixia, Henan

恐龙化石精品

自贡实雷龙足迹
Eubrontes zigongensis

类别：兽脚亚目实雷龙足迹科
产地：内江威远

Classification: Eubrontidae, Theropoda
Locality: Weiyuan, Neijiang

| 2.52亿年 | 2.01亿年 | 1.45亿年 | 0.66亿年 |

| 三叠纪 | 侏罗纪 | 白垩纪 |

何氏极大龙足迹
Gigandipus hei

类别：兽脚亚目极大龙足迹科
产地：内江资中

Classification: Gigandipodidae, Theropoda
Locality: Zizhong, Neijiang

恐龙化石精品

兽脚类足迹和蜥脚类幻迹
footprints of theropods and transmitted prints of sauropods

产地：自贡贡井
Locality: Gongjing, Zigong

| 2.52亿年 | 2.01亿年 | 1.45亿年 | 0.66亿年 |

| 三叠纪 | 侏罗纪 | 白垩纪 |

恐龙化石精品

52亿年	2.01亿年	1.45亿年	0.66亿年
三叠纪	侏罗纪	白垩纪	

兽脚类足迹和蜥脚类幻迹
footprints of theropods and transmitted prints of sauropods

产地：自贡贡井
Locality: Gongjing, Zigong

■ 恐龙化石精品

2.52亿年	2.01亿年	1.45亿年	0.66亿年
三叠纪	侏罗纪	白垩纪	

跷脚龙类足迹
footprints of grallatorids

类别：兽脚亚目跷脚龙足迹科
产地：自贡永年

Classification: Grallatoridae, Theropoda
Locality: Yongnian, Zigong

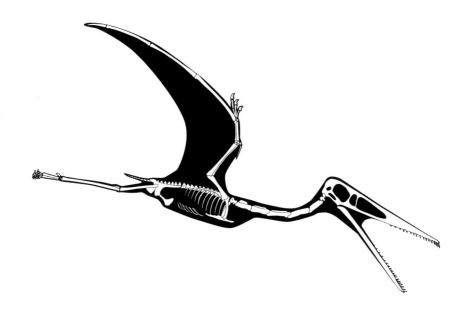
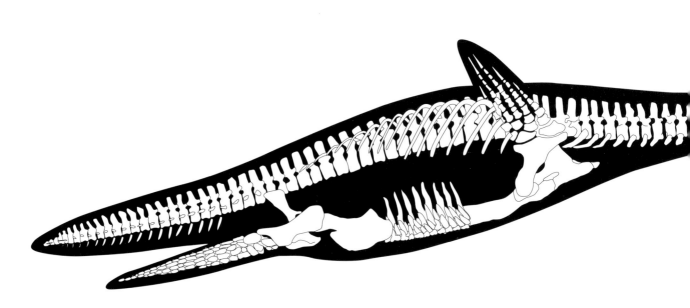

伴生动物化石精品
COEXISTENT VERTEBRATE TREASURES

75 ▶

伴生动物化石精品

大山铺鳞齿鱼
Lepidotes dashanpuensis

类别：半椎鱼目半椎鱼科
产地：自贡大山铺

Classification: Semionotidae, Semionotiformes
Locality: Dashanpu, Zigong

2.52亿年	2.01亿年	1.45亿年	0.66亿年
三叠纪	侏罗纪	白垩纪	

馆藏精品

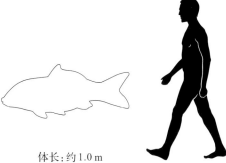

体长：约1.0 m

■ 伴生动物化石精品

刘氏原白鲟
Protopsephurus liui

类别：鲟形目匙吻鲟科
产地：辽宁凌源
Classification: Polyodontidae, Acipenseriformes
Locality: Lingyuan, Liaoning

	2.52亿年	2.01亿年	1.45亿年	0.66亿年
	三叠纪	侏罗纪	白垩纪	

伴生动物化石精品

自贡角齿鱼齿板
dental plate of *Ceratodus zigongensis*

类别：肺鱼目角齿鱼科
产地：自贡大山铺

Classification: Ceratodontidae, Dipnoi
Locality: Dashanpu, Zigong

扁头中国短头鲵
skull of *Sinobra*

类别：片椎目短
产地：自贡大山

Classification: Brac
Locality: Dashanpu

	2.52亿年	2.01亿年	1.45亿年	0.66亿年
	三叠纪	侏罗纪	白垩纪	

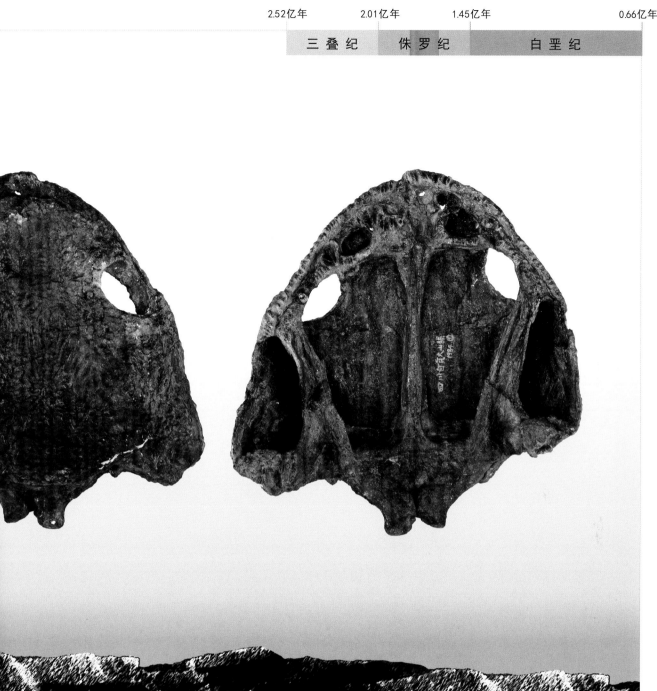

...centicephalus

...nospondylia

81

伴生动物化石精品

自贡巴蜀龟
Bashuchelys zigongensis

类别：伯仲龟亚目巴蜀龟科
产地：自贡大山铺

Classification: Bashuchelyidae, Casichelydia
Locality: Dashanpu, Zigong

大山铺川南龟
Chuannanchelys dashanp

类别：伯仲龟亚目巴蜀龟
产地：自贡大山铺

Classification: Bashuchelyidae,
Locality: Dashanpu, Zigong

自贡天府龟
Tienfuchelys zigongensis

类别：伯仲龟亚目新疆龟科
产地：自贡荣县

Classification: Xinjiangchelyidae, Casichelydia
Locality: Rongxian, Zigong

■ 伴生动物化石精品

杨氏巴蜀龟
Bashuchelys youngi

类别：伯仲龟亚目巴蜀龟科
产地：自贡大山铺

Classification: Bashuchelyidae, Casichelydia
Locality: Dashanpu, Zigong

杨氏巴蜀龟
Bashuchelys youngi

类别：伯仲龟亚目巴蜀龟科
产地：自贡大山铺

Classification: Bashuchelyidae, Casichelydia
Locality: Dashanpu, Zigong

■ 伴生动物化石精品

周氏四川龟
Sichuanchelys chowi

类别：伯仲龟亚目四川龟科
产地：自贡大山铺

Classification: Sichuanchelyidae, Casichelydia
Locality: Dashanpu, Zigong

周氏四川龟
Sichuanchelys chowi

类别：伯仲龟亚目四川龟科
产地：自贡大山铺

Classification: Sichuanchelyidae,
Locality: Dashanpu, Zigong

| 0.66亿年 | 2.52亿年 | 2.01亿年 | 1.45亿年 | 0.66亿年 |

| 白垩纪 | 三叠纪 | 侏罗纪 | 白垩纪 |

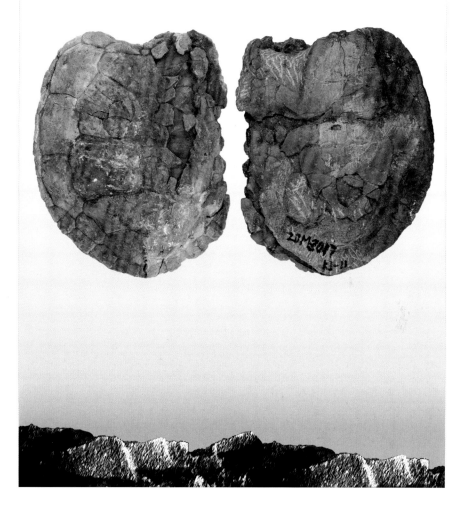

周氏四川龟
Sichuanchelys chowi

类别：伯仲龟亚目四川龟科
产地：自贡大山铺

Classification: Sichuanchelyidae, Casichelydia
Locality: Dashanpu, Zigong

伴生动物化石精品

盐原新疆龟
Protoxinjiangchelys salis

类别：伯仲龟亚目新疆龟科
产地：自贡大山铺

Classification: Xinjiangchelyidae, Casichelydia
Locality: Dashanpu, Zigong

2.52亿年　　　2.01亿年　　　1.45亿年　　　　　　0.66亿年

| 三叠纪 | 侏罗纪 | 白垩纪 |

叶氏香港龟
Hongkongochelys yehi

类别：伯仲龟亚目大贝氏龟科
产地：内江东兴

Classification: Macrobaenidae, Casichelydia
Locality: Dongxing, Neijiang

伴生动物化石精品

2.52亿年	2.01亿年	1.45亿年	0.66亿年
三叠纪	侏罗纪	白垩纪	

放射纹成渝龟
Chengyuchelys radiplicatus

类别：伯仲龟亚目新疆龟科
产地：自贡永嘉

Classification: Xinjiangchelyidae, Casichelydia
Locality: Yongjia, Zigong

宽缘板成渝龟（相似种）
Chengyuchelys cf. *latimarginalis*

类别：伯仲龟亚目新疆龟科
产地：自贡沿滩

Classification: Xinjiangchelyidae, Casichelydia
Locality: Yantan, Zigong

娇小盐都龟
Yanduchelys delicatus

类别：伯仲龟亚目新疆龟科
产地：自贡

Classification: Xinjiangchelyidae, Casichelydia
Locality: Zigong

重庆天府龟（相似种）
Tienfuchelys* cf. *chungkingensis

类别：伯仲龟亚目新疆龟科
产地：自贡汇东

Classification: Xinjiangchelyidae, Casichelydia
Locality: Huidong, Zigong

伴生动物化石精品

| 2.52亿年 | 2.01亿年 | 1.45亿年 | 0.66亿年 |

| 三叠纪 | 侏罗纪 | 白垩纪 |

原新疆龟（相似属）
cf. *Protoxinjiangchelys*

类别：伯仲龟亚目新疆龟科
产地：自贡新民

Classification: Xinjiangchelyidae, Casichelydia
Locality: Xinmin, Zigong

| 2.52亿年 | 2.01亿年 | 1.45亿年 | 0.66亿年 |
| 三叠纪 | 侏罗纪 | 白垩纪 | |

自贡恐龙博物馆 馆藏精品

Casichelydia

95

伴生动物化石精品

| 2.52亿年 | 2.01亿年 | 1.45亿年 | 0.66亿年 |

| 三叠纪 | 侏罗纪 | 白垩纪 |

自贡恐龙博物馆 馆藏精品

汇东四川鳄
Sichuanosuchus huidongensis

类别：原鳄亚目
产地：自贡汇东
Classification: Protosuchia
Locality: Huidong, Zigong

■ 伴生动物化石精品

蜀南孙氏鳄头骨
skull of *Sunosuchus shunanensis*

类别：中鳄亚目 角鳞鳄科
产地：自贡大山铺

Classification: Goniopholididae, Mesoeucrocodylia
Locality: Dashanpu, Zigong

2.52亿年　　2.01亿年　　1.45亿年　　　　　0.66亿年

| 三叠纪 | 侏罗纪 | 白垩纪 |

周氏西蜀鳄头骨
skull of *Hsisosuchus chowi*

类别：中鳄亚目西蜀鳄科
产地：自贡汇东

Classification: Hsisosuchidae, Mesoeucrocodylia
Locality: Huidong, Zigong

伴生动物化石精品

大山铺西蜀鳄头骨
skull of *Hsisosuchus dashanpuensis*

类别：中鳄亚目西蜀鳄科
产地：自贡大山铺

Classification: Hsisosuchidae, Mesoeucrocodylia
Locality: Dashanpu, Zigong

沱江西蜀鳄头骨
skull of *Hsisosuchus tuojiangensis*

类别：中鳄亚目西蜀鳄科
产地：内江凌家场

Classification: Hsisosuchidae, Mesoeucrocodylia
Locality: Lingjiachang, Neijiang

沱江西蜀鳄
Hsisosuchus tuojiangensis

类别：中鳄亚目西蜀鳄科
产地：内江凌家场

Classification: Hsisosuchidae, Mesoeucrocodylia
Locality: Lingjiachang, Neijiang

2.52亿年	2.01亿年	1.45亿年	0.66亿年
三叠纪	侏罗纪	白垩纪	

馆藏精品

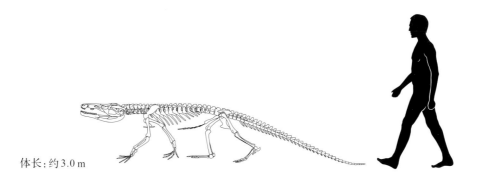

体长：约3.0 m

伴生动物化石精品

自贡璧山上龙
Bishanopliosaurus zigongensis

类别：蛇颈龙亚目拉玛劳龙科
产地：自贡永安

Classification: Rhomaleosauridae, Plesiosauria
Locality: Yongan, Zigong

2.52亿年	2.01亿年	1.45亿年	0.66亿年
三叠纪	侏罗纪	白垩纪	

馆藏精品

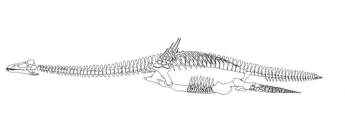

体长：约3.5 m

伴生动物化石精品

贵州鱼龙(未定种)
Guizhouichthyosaurus sp.

类别：鱼龙目萨斯特鱼龙科
产地：贵州关岭
Classification: Shastasauridae, Ichthyosauria
Locality: Guanling, Guizhou

2.52亿年　　　2.01亿年　　　1.45亿年　　　　　　0.66亿年

三叠纪　　侏罗纪　　　白垩纪

馆藏精品

体长：约3.0 m

■ 伴生动物化石精品

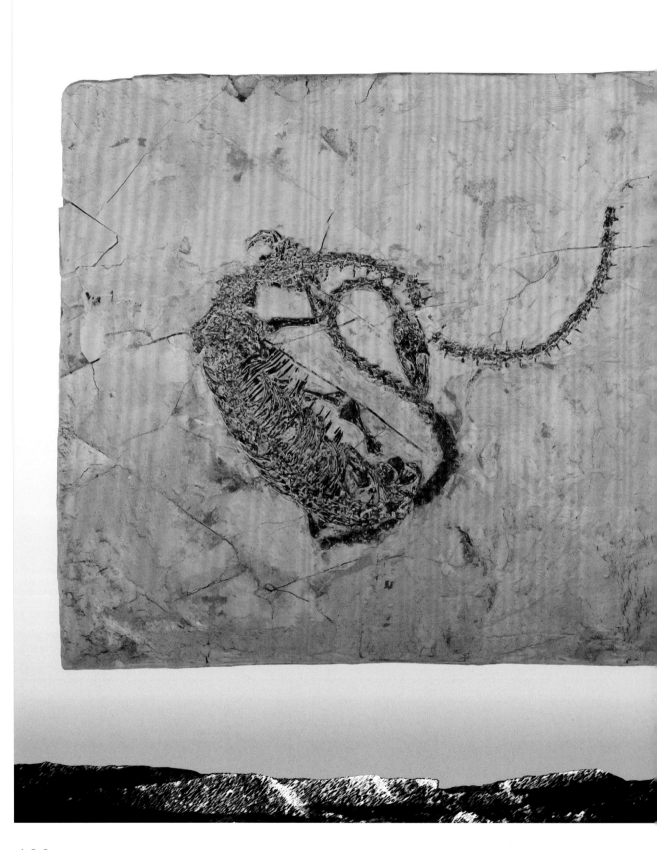

108

.52亿年	2.01亿年	1.45亿年	0.66亿年
三叠纪	侏罗纪	白垩纪	

凌源潜龙
Hyphalosaurus lingyuanensis

类别：离龙目潜龙科
产地：辽宁凌源

Classification: Hyphalosauridae, Choristodera
Locality: Lingyuan, Liaoning

| 伴生动物化石精品

| 2.52亿年 | 2.01亿年 | 1.45亿年 | 0.66亿年 |
| 三叠纪 | 侏罗纪 | 白垩纪 |

长头狭鼻翼龙头骨
skull of *Angustinaripterus longicephalus*

类别：喙嘴龙亚目喙嘴龙科
产地：自贡大山铺

Classification: Rhamphorhynchidae, Rhamphorhynchoidea
Locality: Dashanpu, Zigong

董氏中国翼龙
Sinopterus dongi

类别：翼手龙亚目古神翼龙科
产地：辽宁朝阳

Classification: Tapejaridae, Pterodactyloidea
Locality: Chaoyang, Liaoning

■ 伴生动物化石精品

川南多齿兽头骨
skull of *Polistodon chuannanensis*

类别：兽齿亚目三列齿兽科
产地：自贡大山铺
Classification: Tritylodontidae, Theriodontia
Locality: Dashanpu, Zigong

爬兽（未定种）头骨
skull of *Repenomamus* sp.

类别：真三尖齿兽目爬兽科
产地：辽宁北票

Classification: Repenomamidae, Eutriconodonta
Locality: Beipiao, Liaoning

伴生动物化石精品

孤反鸟（未定种）
Monoenantiornis sp.

类别：反鸟目反鸟科
产地：辽宁凌源

Classification: Enantiornithidae, Enantiornithiformes
Locality: Lingyuan, Liaoning

| 2.52亿年 | 2.01亿年 | 1.45亿年 | 0.66亿年 |

| 三叠纪 | 侏罗纪 | 白垩纪 |

朝阳鸟（未定种）
Chaoyangia sp.

类别：朝阳鸟目朝阳鸟科
产地：辽宁朝阳

Classification: Chaoyangidae, Chaoyangiformes
Locality: Chaoyang, Liaoning

■ 伴生动物化石精品

手兽足迹（未定种）
Chirotherium sp.

类别：手兽足迹科
产地：云南会泽

Classification: Chirotheriidae
Locality: Huize, Yunnan

自贡恐龙博物馆地理位置图
LOCATION SKETCH MAP OF ZIGONG DINOSAUR MUSEUM

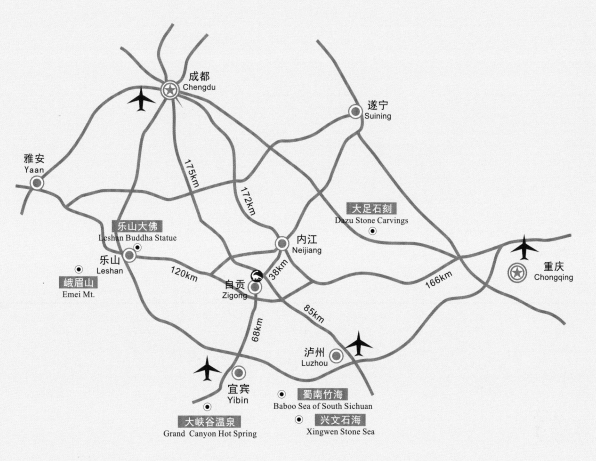

地址：中国四川省自贡市大安区燊海井路四段268号
Add: No.268, Section 4, Shenhaijing Rd., Daan, Zigong, Sichuan 643013, P.R.China
电话(Tel): +86-813-5801000　　　　传真(Fax): +86-813-5801233
http://www.zdm.cn　　E-mail:zdmoffice@126.com　　邮编：643013